GRASSHOPPERS and CRICKETS

By the Author

SHORT-HORNED
GRASSHOPPER

Dorothy Childs Hogner

GRASSHOPPERS and CRICKETS

Illustrated by Nils Hogner

Thomas Y. Crowell Company, New York

For Susan

Contents

GRASSHOPPERS and CRICKETS

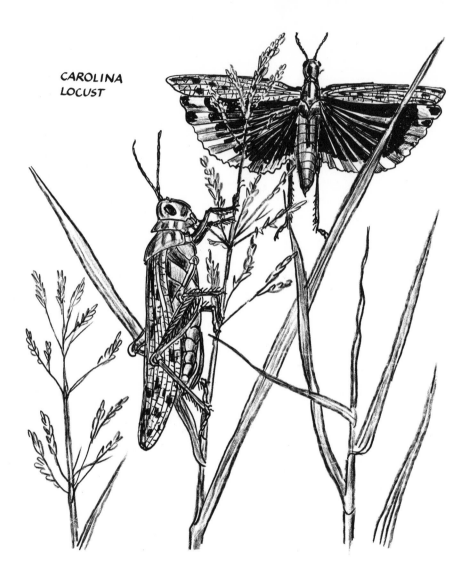

CAROLINA
LOCUST

The Grasshopper at Home

Where does one look for grasshoppers? The answer is in their name, for they live everywhere, all over the world, where there is *grass*. You can also find certain kinds in the desert, even in Death Valley.

The common, winged kinds are easy to find because they do not make nests or dig burrows to hide in. They live in the open, with no roofs over their heads.

In the summertime, look for them on the lawn. Lie down in a meadow and you will see dozens of them, hopping in the tall grass. They are common in cow pastures and where horses and ponies graze.

If you live on the Great Plains in the West, you may chance to see millions of them, all at one

1

time, flying into a field of grain. We hope you don't. When swarms of grasshoppers invade a field, they eat so much that they ruin the crop.

Certain wingless grasshoppers, which look more like crickets, are at home under stones and in rubbish heaps, or in sandy soil.

Is a Grasshopper an Insect?

Some people think that just any little crawling thing is an insect. But not all of them are. Here is

2

how to tell an insect at a glance. Count the legs. All full-grown insects have six legs.

Now compare the number of legs on a spider with those of a grasshopper. A spider has eight legs, which proves at once that it is not an insect. A spider belongs to a class of eight-legged creatures which also includes scorpions, mites, and ticks, among others.

A grasshopper, which is a real insect, has only six legs.

Mightiest Jumper

The way in which a grasshopper uses its legs is told in the second half of its name. For what does a grasshopper do in the grass? Most often, it hops around.

In a common grasshopper there is a great difference in size between the fore and hind pairs of legs. The two front pair are short and slender, while the hind pair are long, stout and very powerful. In fact, the hind legs of a grasshopper have the most powerful muscles, in proportion to size, of any creatures in the animal kingdom, except fleas. A flea is probably the most powerful jumper of animals, in proportion to size. The muscles with which a clam closes its shell are more powerful than a grasshopper's leg muscles but a grasshopper works its muscles faster. A grass-

FLEA

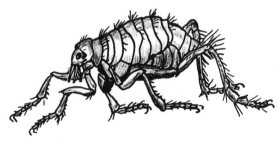

hopper uses only its hind legs for jumping and it is a mighty jumper.

A grasshopper can jump twenty times its own length, broad jumping, and ten times its length, high jumping. If a man could jump with equal power, in proportion to his height, he would think nothing of leaping over his own house.

But if a man could leap that high, he would be killed on hitting the ground, because of his weight. The grasshopper, on the other hand, can control its flight with its wings. But it does not always do so. It may turn over and over and crash-land on its feet, its side, or even hind side foremost, without harm to its body, because it is so light.

This great jumper in action is a wonderful sight. It gets set by drawing the shins of its hind legs against the thighs at a sharp angle. (Note that it

bends its jumping legs in the opposite direction from which we bend our legs.) Then it rears up and pushes off.

You can make a grasshopper jump by touching the tail end of its belly. Left to itself, a grasshopper does not always jump. It has amazing control over its leg muscles and when it chooses, it hops slowly, or just walks.

Inside Out

The skeleton of a grasshopper is very different from ours. Our skeleton is, as it were, inside of us. Skin and muscles cover it.

With a grasshopper, quite the opposite is true. The skeleton is the first thing one sees when one looks at a grasshopper because it is outside of the insides. It is a kind of tube which covers the muscles and the blood vessels, as well as all the organs.

This covering is made of a horny material called chitin. Though not as hard as bone, it protects the grasshopper well, for it is formed in ringlike, overlapping sections. The body wall between is soft, which enables the grasshopper to bend and move its body.

The body itself has three divisions, the head, the middle or thorax, and the abdomen or belly.

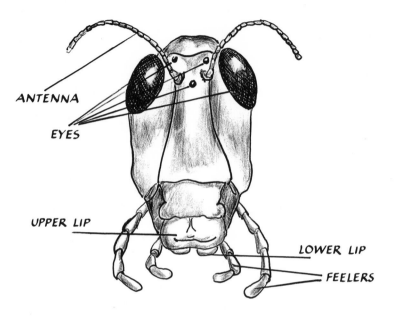

ANTENNA

EYES

UPPER LIP

LOWER LIP

FEELERS

The Head

The first section, the head, bears a pair of
feelers, and the eyes and the mouth. The feelers
and antennae are made of jointed rings of chitin,
the same material as the body covering, so the
grasshopper can wave its feelers every which
way.

Just by looking at a common grasshopper's feelers, you can tell to which of two grasshopper families it belongs. If the feelers are longer than the body, it belongs to the long-horned grasshopper family. If the feelers are shorter than the body, it is a member of the short-horned grasshopper family.

Try to catch either a long-horned or a short-horned grasshopper by hand and you will soon

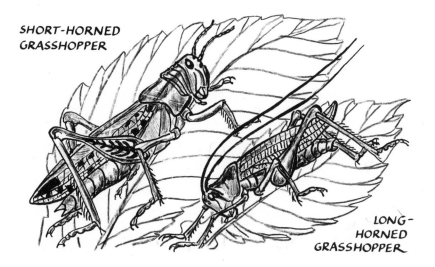

SHORT-HORNED
GRASSHOPPER

LONG-
HORNED
GRASSHOPPER

learn that a grasshopper can see from several directions at once. This is because it has five eyes.

Three of its eyes are so small that they are hard to see. One of these is located on the inside edge of each of the two big eyes, and the third is in or near the middle of the face. The three little eyes see only light, or at most, an object nearby.

The two big prominent eyes, one on each side of the head, are extraordinary. They are not single, like the little eyes. These big eyes are made up of hundreds of six-sided pieces, each of which is a small separate eye in itself. A moving object is seen by every one of the little eyes within the big eyes. That is why a grasshopper can see your hand in time to escape, unless you are very quick.

A grasshopper's mouth is complicated, too. Insects, such as bees and butterflies, which suck their food, have sucking mouths. The mouth of a

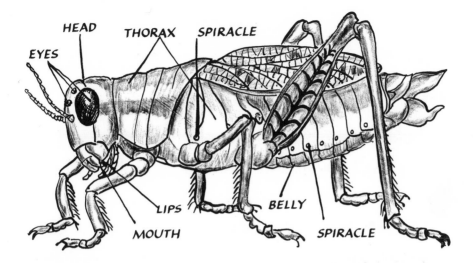

HEAD THORAX SPIRACLE
EYES
LIPS
MOUTH BELLY
SPIRACLE

grasshopper, a chewing insect, has several parts: a tonguelike organ, a pair of lips, and two pairs of jaws. The single upper lip and the lower lip, which has two flaps and two feelers, hold the food between the jaws. Two of the jaws are hard, with toothlike edges, and do most of the grinding.

Thus does a grasshopper chew the grass, clover, and other greens upon which it feeds.

11

The Middle
and the Abdomen

Joined to the head is the middle or thorax. The thorax itself consists of three parts, each of which bears one pair of legs.

The two hind parts of this middle section also bear the wings which many grasshoppers have when full grown. The first pair of wings are tough. They are called the wing covers because when the grasshopper is not in flight, they lie over the thin, fragile, second pair. The second pair are folded, when at rest, in pleats.

Each wing has many veins which help to make the wing strong. But each kind of grasshopper has a different arrangement of veins in its wings.

Joined to the hind end of the thorax is the third

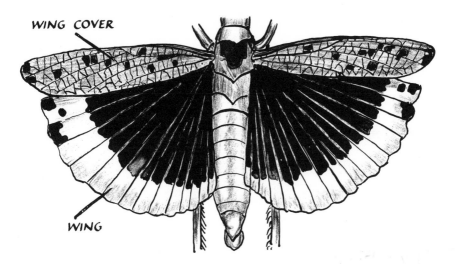

WING COVER

WING

section of the body, the abdomen or belly. At the tail end of the belly of the female grasshopper is the egg depositor.

How a Grasshopper Hears, Smells, and Breathes

Do not look for ears or a nose on a grasshopper's head because they are not there. In fact,

13

a grasshopper has no real nose, such as animals have. Its organs of smell are located on its feelers, but it does not breathe with its "smeller" as we do. It breathes through holes, called spiracles, in its sides. Two pairs of spiracles are located on the wing-bearing parts of the middle section of the body, and a pair on each section of the belly, one on each side.

The ears of a short-horned grasshopper are located on the front ring of the abdomen, one on each side. Each eardrum is stretched between a small ring of chitin.

The eardrums of a long-horned grasshopper are on the lower forelegs.

SHORT-HORNED GRASSHOPPER

The Song of a Grasshopper

A good way to locate grasshoppers in late summer and fall is to stop and listen. At this time of year, the males of many kinds of grasshoppers "sing" and each kind has a special note of its own.

A grasshopper does not sing in the manner of a canary. It produces music more as a violin does. Young grasshoppers cannot sing because they are wingless, and a grasshopper's wings are its violin.

Certain short-horned grasshoppers sing by rubbing the lower part of their hind legs against the rubbing parts of their wing covers. The row of small hairlike prickers on each leg thus acts as a bow, and the wings as a violin.

Other short-horned grasshoppers make a crackling sound by rubbing their wing covers

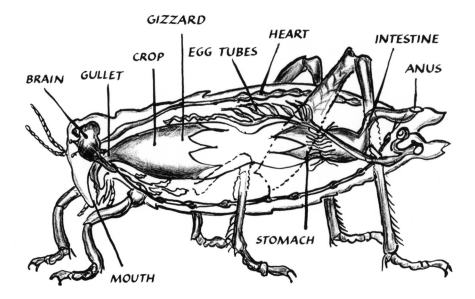

BRAIN   GULLET   CROP   GIZZARD   EGG TUBES   HEART   INTESTINE   ANUS

MOUTH   STOMACH

against their hind wings. They do this while fly-ing.

The males of long-horned grasshoppers sing by rubbing their wing covers together until they vibrate.

The arrangement of the veins on a female grass-hopper's wings is such that, in most kinds, she cannot make sounds.

16

A Grasshopper on the Inside

The heart of a grasshopper is a long tube divided into chambers and open at both ends. It lies under the muscles of the back.

The brain is in the head.

The lungs are a system of air sacs which serve a double purpose. They take air in through the breathing holes in the grasshopper's sides, and bring life-giving oxygen to every part of the body. Then the lungs gather up the harmful carbon dioxide which has formed and pass it back out through the breathing holes.

Down the center of the body, from end to end, runs the alimentary canal through which food passes and is digested. Food goes from the mouth to the gullet and into the crop where it may be stored temporarily; from there it goes into the muscular grinder, the gizzard, and on into the stomach which pours out digestive juices. The food passes through the intestines, and the wastes go out the opening at the end, the anus.

The sex organs are in the third, the abdominal, section of the body. Grasshoppers reproduce by laying eggs, and in the female are two sets of egg tubes where the eggs are formed.

A Grasshopper's Birthday

Egg-laying time for most grasshoppers is the late summer or early fall. (A few kinds may have two broods a year, a spring-born brood and a late summer one.)

The mother grasshoppers—the short-horned kind—dig holes in the ground, using their egg

depositor as a digging tool. Some dig holes in
old stumps, others in old wooden fences. The
mother deposits an oval mass of two to three
dozen eggs in the hole and covers the eggs with

19

a gummy material. During egg-laying time, she will deposit usually two, sometimes three, egg pods. After she deposits her eggs, off she goes, giving not a thought to the offspring she will never see.

When winter comes, the eggs remain snug in the egg holes, but the old grasshoppers (with the exception of the pygmy short-horned, and certain other species which live in the South) do not live over winter. Most of the common kinds are gone when the little grasshoppers break out of the eggs in the spring.

You may find both tiny grasshoppers and big ones hopping around at the same time because some hatch early and some late.

The change from a baby to a full-grown grasshopper is not startling, like that of some insects. A moth, for example, comes out of its egg in the

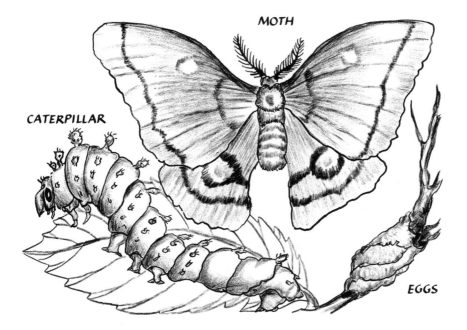

MOTH

CATERPILLAR

EGGS

form of a larva, a little caterpillar that looks more like a worm than a moth. When full grown, the larva spins a cocoon; some time later, out of the cocoon comes a beautiful six-legged, winged moth, a creature that is entirely different from the larva.

Not so with a grasshopper. The baby grasshopper, called a *nymph*, is so small that few people would notice it, but it looks enough like its parents for anyone to see that it is a grasshopper. It has no wings, but it has wing buds, six legs, and a very large head.

The change from a nymph to a full-grown grasshopper comes about little by little. The nymph gorges itself on the same green food that its parents eat. As it grows, its outside skeleton, softer at first than that of a grown grasshopper, stretches, but only just so far. Finally the little grasshopper gets too big for its body covering and the covering splits down the back.

Meanwhile, a new body covering has been forming under the old, and soon the little nymph, in a new jacket, crawls out. The split skeleton it leaves behind is a transparent hollow form of a little grasshopper, complete, even to legs and

NYMPHS

feelers. Look carefully where grasshoppers are common and you may find one.

The new covering soon hardens with chitin, and so, as the nymph continues to grow, it must shed, not once but several (usually four or five) times. This is called molting.

At the last molt, the wings are complete and the grasshopper can fly.

Short-Horned Grasshoppers, the True Locusts

There are many different kinds of short-horned grasshoppers, which are also correctly called locusts.

Several kinds with red legs, including the one called the red-legged grasshopper, are common nearly everywhere in the United States where

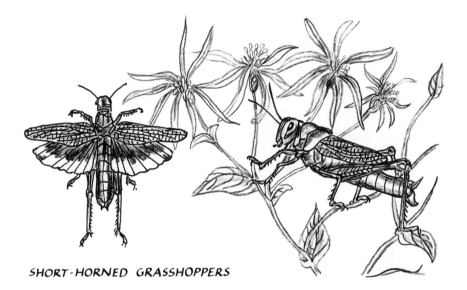

SHORT-HORNED GRASSHOPPERS

there is grass. One place the red-legged grass-hopper is not found is the dry grassy land in the mountains of the West.

In the Rocky Mountain region lives the western migratory, or Rocky Mountain, locust. The Rocky Mountain locust does not, however, always stay in the mountains. It may be found as far away as the Mississippi Valley.

Both the red-legged and the Rocky Mountain locusts are strong fliers.

A short-horned grasshopper called the lesser migratory locust is common in the northern parts of the United States.

A common short-horned grasshopper in the East is the clouded locust. In color, it is earthy brown with spots. The male makes a crackling sound with its wings in flight.

All of these are quite stoutly built. Certain slender short-horned locusts live in the South and the West.

The Carolina locust is common in grassy land not only in North and South Carolina, but in many places in the United States and southern Canada. It is the biggest of the short-horned grasshoppers and may reach nearly two inches in length. The color of its body varies with the

CLOUDED
LOCUST

ground upon which it lives, so that it may be brown or reddish or yellowish. There is a broad yellow edge on its black hind wings. It is so fond of alighting on roads that it is often called the "dusty-roads grasshopper."

Long-Horned Grasshoppers

Most familiar of the long-horned grasshoppers is the large, usually apple-green, Katydid. (Some time you may see a Katydid that is pink, for a few are that color.)

Who has not heard the male Katydid rub its angular wing covers together until they vibrate with the Katydid's loud, rasping song, "Katy-did, Katy-didn't, she did, she didn't!" Whether the male is singing to or about Mrs. Katydid is hard

KATYDID

to say. Mrs. Katydid herself makes no reply because she cannot make music.

Mrs. Katydid is one kind of grasshopper that does not use her prominent, sword-shaped egg depositor as a digging tool because she does not lay her eggs under the surface of the ground. In fact, she is very casual about where she does lay them. Sometimes she chooses a twig; she roughens the surface with her mouth, then deposits her eggs in a double row. At other times she may lay on any old thing, such as a towel hanging on the wash line. In the South she frequently lays her eggs on the edge of a leaf.

She may deposit only two eggs at one place, or as many as thirty. Before winter comes she will have laid a total of a hundred to a hundred and fifty eggs. In the South she may lay even more, for there she often lays eggs for two hatchings.

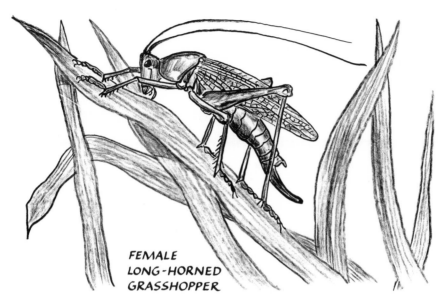

FEMALE
LONG-HORNED
GRASSHOPPER

Almost as familiar as the Katydids are the meadow grasshoppers. These small to medium-sized, delicately built insects are common in wet fields and swampy places. Some lay their eggs under the surface of the ground, others lay in the pithy stems of plants.

Most long-horned grasshoppers eat green food. A few also eat other insects.

31

Cricketlike
Long-Horned Grasshoppers

There is a certain kind of long-horned grass-hopper that many people mistake for the cricket. These grasshoppers look like crickets and they are even called crickets, but they are real grass-hoppers. The best known of these are the "Mormon cricket" and the "sand cricket." You can tell them apart from the common black cricket at a glance because neither of them has wings.

The Mormon cricket is a large, black, long-horned, wingless grasshopper most commonly found in the high mountains of the West. The female has a long egg depositor with which she lays her eggs deep under the surface of the ground.

The sand cricket is a long-horned, wingless

MORMON CRICKET

grasshopper which lives in sandy soil and under rocks on the West coast. It is a big insect with a big head.

The Harm Grasshoppers Do

Every few years, certain kinds of short-horned grasshoppers—the true locusts—do great damage. This happens when they hatch out in unusually large numbers. There is, of course, not enough food for all of them where they hatch. So, unlike most kinds of grasshoppers, which live out their lives near where they are born, the locusts migrate, looking for food. The young ones hop along. Those which hatched early and have their wings, fly. They invade fields of wheat, they go after cotton, corn, rye, oats, barley and alfalfa, to mention some of the food they like to eat.

These are the grasshoppers which, since the days when Moses in the Bible foretold a plague of locusts, have continued to do great harm to crops. All over the grass-covered world, from

Egypt to Russia and India, from southern Europe to Africa, from South America to our own states, short-horned grasshoppers have brought destruction.

In years when Mormon crickets are abundant, they, too, are a plague. Being wingless, they walk and hop down from their dry mountain homes into the green valleys and onto the plateaus. They may completely destroy a garden of vegetables. They devour sugar beets and eat the seed heads off wheat.

However, because much of the land they invade is not crop but range land, Mormon crickets do most damage to the range where cattle, sheep and horses graze. They are known to eat 250 kinds of range plants. And because they like best to feed on the flower and seed parts, they not only take fodder intended for livestock, but they

35

prevent range grasses from reseeding themselves.

The year 1938 saw one of the worst outbreaks. In that year Mormon crickets damaged range land in Colorado, Idaho, Montana, Nebraska, Nevada, North Dakota, South Dakota, Oregon, Utah, Washington, Wyoming and California.

Another glutton in the long-horned grass-hopper family is the Katydid. Fortunately, Katy-dids never appear in such numbers as do the Mormon crickets and the locusts. But even a few Katydids can ruin a rose bed. They like nothing better than to sit on a rosebud and chew holes in it.

Enemies of Grasshoppers

Among the living creatures that feed on grass-hoppers are birds, other insects, and some ani-

mals such as cats. But, as one can easily imagine,
the worst enemy of grasshoppers is man, who has
waged a fight against them for centuries.

Nearly a hundred years ago in this country, the Congress of the United States recognized how serious a problem grasshoppers can be. At that time the Rocky Mountain locusts caused a calamity. They appeared in hordes on the plains east of the Rocky Mountains in Wyoming, Colorado and Montana; they migrated into Texas and down into the Mississippi Valley, devouring crops wherever they stopped in their flight. In three years they did 200 million dollars' worth of damage. As a result, Congress set up an insect commission, whose first job was to plan what to do about grasshoppers.

Farmers began destroying grasshoppers with poisoned bran bait. Today, the common way to stop these insects from ruining farm and range crops is to spray them, often from airplanes, with poison sprays.

Cousins of the Grasshoppers

One of the few grasshopperlike insects which is a friend to man is the praying mantis. This cousin of the grasshopper is named for its habit of standing still, with its forelegs raised, as if in prayer. In this pose, the mantis stands ready to pounce upon any unwary insect that comes

PRAYING
MANTIS

close. For while most grasshoppers are vegetarians, praying mantes are strictly meat eaters. A praying mantis takes and eats every insect within reach of its grasp.

Another cousin of the grasshopper, the cockroach, likes to live in houses and feed on crumbs and garbage left about by untidy people. It is a pest.

Most interesting of the cousins of the grasshopper are the crickets.

COCKROACH

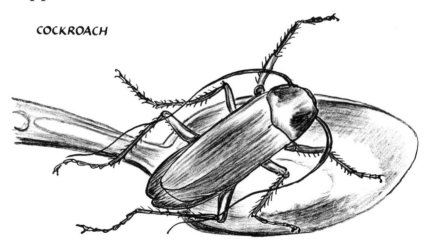

When Is a Cricket a Real Cricket?

Since some grasshoppers are called crickets, how does one know when one finds a real cricket? It is easy. Just look at the insect's wing covers. The wing covers of a grasshopper meet more or less in a peak, the wings lie at rest on a slope. Crickets' wing covers lie flat on the back and fold down at the sides, somewhat like a tablecloth.

Crickets have long feelers. The egg depositor of the female is shaped like a spear.

Among the several kinds of crickets is the mole cricket, which is especially fitted for life underground. Like a real mole, the mole cricket has stout forelegs and big forefeet, useful for digging its burrows. Where mole crickets appear in num-

41

bers, they are a pest because they feed on the roots of plants.

The tree cricket is so named because it often lives in trees. The common tree cricket in the East is the whitish-green, snowy tree cricket; it also lives on shrubs and plants. It is not chubby, like the common cricket, but fragile-looking, with delicate, transparent wings. It has good jumping legs.

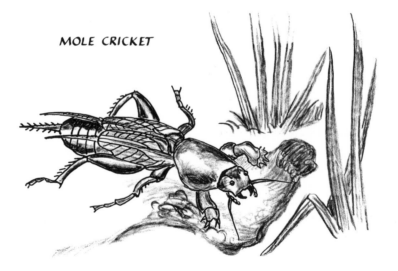

MOLE CRICKET

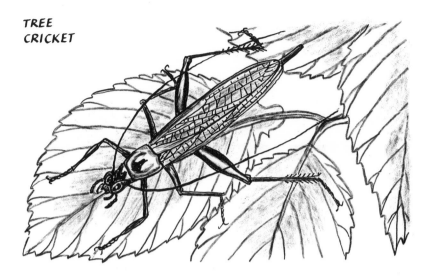

The female snowy tree cricket sometimes lays her eggs in raspberry canes, which kills the canes above the egg hole.

The Friendly Common Cricket

The true cricket is the familiar, stubby-built, black insect that likes to move into peoples'

43

FEMALE CRICKET (LEFT) LISTENING TO SONG OF MALE

houses and sit by the fire and sing. It thus brings cheer to a household and, some people say, good luck.

The violin of this cricket, like that of a winged grasshopper, is its wings. There are ridges on the main vein of the wing covers of a male cricket and there is a hard spot near the end of the vein. In order to sing, he raises his wing covers and, holding them thus, rubs them, one against the other, so that the hard spot on one goes back and forth against the ridges on the opposite wing.

Because the female cricket has plain veins on her wings, she cannot sing, but she can hear her mate's song. Like long-horned grasshoppers, the crickets have little, drumlike surfaces on the lower parts of their forelegs, and these are their ears.

Cricket Music in China

For hundreds of years, people in China have kept crickets, much as we keep canaries, just so they could enjoy their music.

People in old China used little bamboo traps to catch crickets. A wealthy person might carry an ivory trap, adorned with a tiny dragon. Cricket hunting took place after dark. A candle or a charcoal-burning iron basket was carried, and the light attracted crickets from their hiding places.

Once a cricket was caught, the Chinese would keep it in a small cage, usually made of bamboo or wood, although a rich lady might keep her cricket in a cage of gold.

In summer, the cricket pet was usually put in a little covered pottery jar, so it would keep cool. In each jar the owner set tiny dishes to hold food and water, and sometimes a little clay bed on which the cricket could sleep.

In winter, it might be moved into a gourd that had been especially grown to be a cricket house. First, the flower of the gourd was placed in a mold, carved on the inside. When the flower faded and the fruit formed, the fruit was squeezed into the shape of the mold. The design on the inside of the mold was thus pressed into

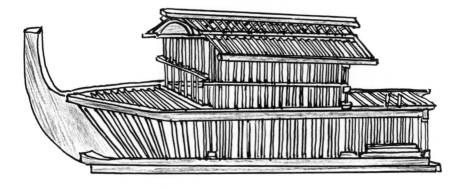

OLD CRICKET CAGE

the soft skin of the gourd. After the gourd dried, the top was sliced off and the insides removed. A cover, to keep the cricket in the gourd, might be made of coconut, sandalwood, costly jade or ivory; it was carved with an open-work design to let in air.

When the owner cleaned the cricket cage with a small brush, he held his pet cricket under a tiny wire net to keep the cricket from escaping. The crickets were encouraged to sing by a tickler made of the whiskers of a rat or hare, fastened in a reed, bone or ivory handle.

In old China it was the custom for people to take pet crickets about with them, much as we take dogs on leash. The cricket might hang in a cage from a belt; or, if the weather turned cold, the owner would hang the cricket cage under his shirt to keep the cricket warm. Thus, while walking down a street, one might hear the cheery sound of crickets chirping, as people went about their business. On coming into a Chinese house, one was often greeted by a cricket orchestra, for many people kept more than one cricket at a time.

A popular tree cricket in China is called the Golden Bell because its note is said to resemble the tinkle of a little bell. And because the Golden Bell male never sings unless there is a female nearby, a female Golden Bell is also kept in the same cage.

Other female crickets caught in China, unless kept for breeding, are fed to the birds.

Cricket Fights

A cricket chirping by the hearth sounds so peaceful and friendly that it is surprising to learn that crickets are savage fighters. Each cricket likes a home of its own and will fight, even to death, to keep other crickets from entering its chosen hole. In the old days in China, men believed that a brave fighting cricket was really a man, a great hero, who had come back to life in the form of a cricket. Fighting crickets were kept, as bulls are in Spain, for tournaments.

A good fighter was sometimes valued as much

49

as a horse and was fed special food. There were heavyweight, middleweight and lightweight cricket matches and, before a fight, each cricket was weighed in on a tiny scale. The cruel fight, which took place in a pottery jar, usually ended by the victor tearing his opponent apart.

Your Own Cricket Cage

For several hundred years the people of Japan, like those of China, have kept crickets as pets. Today, in Tokyo, cricket merchants set up stalls at festivals and sell several kinds of Japanese crickets and little cages to keep them in.

Shops in America that sell goods from Japan usually carry Japanese cricket cages. If you buy one, be sure that it is a *live* cricket cage. Some stores sell toy cricket cages and a toy cricket to sit inside. Many of these cages have bars spaced so far apart that even a full-grown cricket can crawl right out.

It is possible to cover such a cage with netting but this is not practical.

A pretty, antique live cricket cage is shown on page 47. In this the bars are only one eighth of an inch apart, so that even a young and quite small cricket is unable to escape.

If you cannot find a live cricket cage in a store near where you live, a glass terrarium, such as you can buy in the five-and-dime store, makes a dandy substitute. A terrarium, of course, is made for growing plants indoors. It need be only big enough for you to put your hand in. When living creatures are kept in a terrarium, it is called a vivarium.

An aquarium, without water naturally, may also be used as a cricket home.

You must screen the top of a terrarium or your cricket will soon hop out.

To Find a Cricket

In late summer or fall, you will very likely find a quite large cricket in your house. If not, look around outside. You may come upon one in day-time, peeking out from a crack in the sidewalk or from under a porch. You may see one crawling out from under a stone or an old log. A good place to look is in a hayfield or along a roadside. If you don't find one hopping at your feet, look under the edge of a pile of dead grass, a haycock, a haybale, or try the haymow in the barn.

Also in late summer and fall, it is fun to hunt crickets after sunset because crickets like best to sing at dusk and in the dark. Take a flashlight, go outdoors, sit down and listen. When you hear a chirp, switch on your light and move slowly toward the sound. This is not as easy as you may think, for a cricket is a ventriloquist. That is, it can throw its voice, so that, while it seems to be on your left, it may actually be on your right.

If it stops singing, just stand still until it begins again. When you catch sight of the insect, move very slowly, or it will hop off. Then quickly cup your hand over it and hold it gently—then you won't break off the feelers and legs.

If you chance to catch a cricket that is not singing, look to see if it has a long egg depositor at the end of its abdomen. If it has, it is, of course, a female. Let it go and try again; only a male sings.

The Cricket in a Cage

It is easy to care for a cricket in a live cricket cage, such as shown on page 47. Because the whole top of this cage lifts off the cage's floor, you can set the top on a table, with the live cricket still in it. The table acts as a temporary floor while you brush out the old food and droppings with a small artist's paintbrush and put in fresh food and water.

If you keep a cricket in a terrarium or aquarium, place the insect under a tea strainer while you clean out its home. If it should get loose, it may eat holes in your clothing.

A bottle cap makes a good cricket drinking dish. Keep it filled with clean water.

As to food, a cricket likes both fresh and decaying vegetable matter, and dead insects, such

as flies. It eats many kinds of plants, including
lettuce, and is especially fond of soft fruits, such
as bananas; it also likes soft cheese. Feed it only
part of a lettuce leaf and a small piece (about the
size of a nickel) of cheese or banana at a time.

If you put two crickets of the same sex in the same cricket cage, chances are that, by the next morning, there will be only one. Two females, or two males, may fight, and the victor eats the loser.

In the open, most old crickets do not live over winter. But a few, particularly of those which find shelter in a building, survive cold weather. So bring your cricket cage indoors when frost comes. Then, if you give your cricket good care, it will very likely keep healthy and sing for you all winter long.

You may, of course, keep a grasshopper or a Katydid in the cage, instead of a cricket. Grasshoppers are very fond of clover, so if you live in the country, pick fresh clover leaves for your pet every day. You can plant the floor of the terrarium with clover plants, dug in a field. As soon as the grasshopper has chewed off all the leaves, remove the old clover and plant a fresh clump.

In winter, feed grasshoppers lettuce or other greens from the market.

Whether you keep a cricket or grasshopper in a cage, or merely watch them hopping and fiddling in the field, you are certain to find them interesting members of the big insect world.

INDEX

About the Author and the Artist

Dorothy Childs Hogner is a Connecticut Yankee, born in Manhattan. The daughter of a doctor, she lived the first year of her life in New York. Then her family moved to an old white clapboard house on a hundred-acre farm in Connecticut.

Mrs. Hogner attended Wellesley College in Massachusetts, Parsons Art School in New York, and was graduated from the University of New Mexico. She is the author of many books for children, several of which are illustrated by her husband, Nils Hogner.

Mr. Hogner, who illustrated *Grasshoppers and Crickets*, is primarily a mural painter. One of his historical murals is in the high school in Litchfield, Connecticut, and his Memorial to the Four Chaplains may be seen at Temple University in Philadelphia.

The Hogners' winters are spent in their New York City apartment and their summers on a herb farm in Connecticut, where they raise everything from basil to sweet cicely. It was on their farm that they captured and raised the grasshoppers used by Mr. Hogner as models for the drawings in *Grasshoppers and Crickets*.